AF262942

5-Minute Genius Stories: Albert Einstein © 2026 Lucky Cat Publishing Ltd
Illustrations © 2026 Sally Walker

Union Square Kids
Hachette Book Group
1290 Avenue of the Americas, New York, NY 10104
unionsquareandco.com
@unionsqandco

First Edition: June 2026

Union Square Kids is an imprint of Grand Central Publishing, a division of Hachette Book Group, Inc.
The Union Square Kids name and logo are registered trademarks of Hachette Book Group, Inc.

The publisher is not responsible for websites (or their content) that are not owned by the publisher.

Union Square Kids books may be purchased in bulk for business, educational, or promotional use.
For information, please contact your local bookseller or the Hachette Book Group Special Markets Department at special.markets@hbgusa.com.

Text by Wil Mara
Book design by Ashtyn Botterill
Edited by Rosie Neave

Library of Congress Cataloging-in-Publication Data has been applied for.

ISBN 978-1-4549-6174-1

Printed in Guangdong, China
TLF 02/26

2 4 6 8 10 9 7 5 3 1

For Jenna—W.M.

For Sadie and Elliott, my universe—S.W.

5 Minute Genius Stories

ALBERT EINSTEIN

Written by Wil Mara

Illustrated by Sally Walker

How to use this book

In this book you'll find **ten genius stories** to read, each one just **5 minutes long**.

At the end of each story, explore an informative **"all about"** spread.

Want to learn more? Turn to the back of the book to discover a **timeline of key events**.

union
square
kids

NEW YORK

WHO WAS ALBERT EINSTEIN?

Albert Einstein is best known as one of the greatest physicists in history. His groundbreaking ideas forever changed the way we think about the Universe. Follow his path to becoming a physics icon, as well as the very first celebrity scientist.

A compass helps Albert discover his passion for finding out how the Universe works.

Page 8

A family friend brings Albert a set of science books—sparking a lifelong fascination with physics.

Page 18

Albert's confidence in his growing genius makes him a challenging student.

Page 28

Albert questions the then-established theory that light travels through the air as a wave.

Page 38

MOTION COMMOTION

Albert realizes that tiny atoms play a huge part in some mysterious movements.

Page 48

ANNALS OF PHYSICS

Albert demonstrates that space and time are connected.

Page 58

THE MOST FAMOUS EQUATION

Albert has the idea that everything with mass also has energy, and describes it: $E=mc^2$.

Page 68

THE MIRACLE YEAR

Albert shows that large objects have a powerful gravitational effect on nearby objects.

Page 78

THE BIGGEST PRIZE OF ALL

Albert is recognized for his revolutionary work when he is awarded the Nobel Prize for Physics.

Page 88

ONE MORE BIG IDEA

Albert tries to come up with a single explanation for all the forces that control the Universe.

Page 98

WHICH 5-MINUTE GENIUS STORY WILL YOU READ TODAY?

HIDDEN POWERS
The Mysteries of Science

Albert Einstein was born in the German town of Ulm on March 14, 1879.

The next year, his family moved to the busy city of Munich. Albert's father and uncle started a company that made electrical equipment. Then Albert's sister Maria was born.

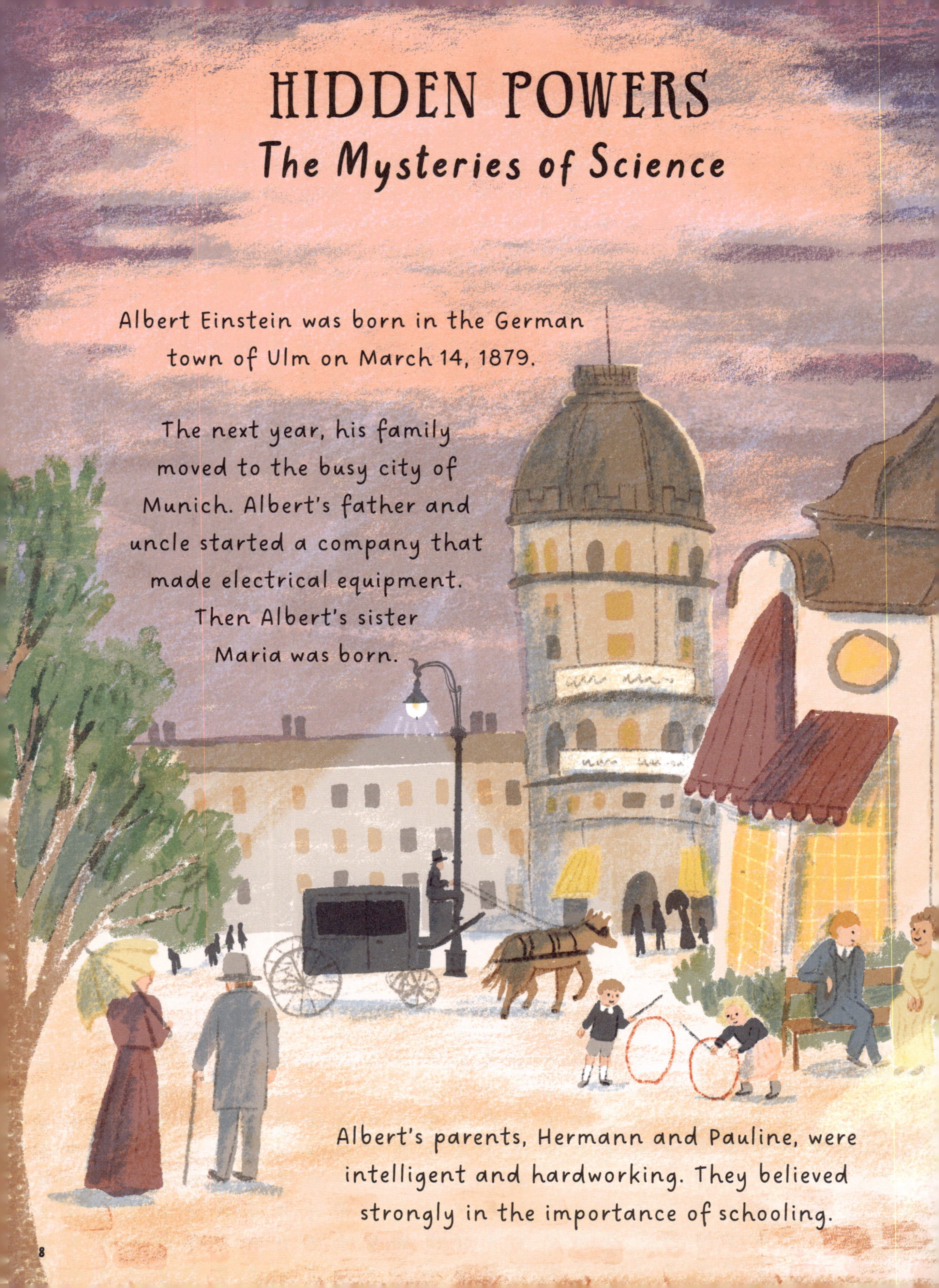

Albert's parents, Hermann and Pauline, were intelligent and hardworking. They believed strongly in the importance of schooling.

Unfortunately, little Albert had his parents very worried. It took him longer than other children to begin speaking.

There was certainly no reason yet for Albert's mom and dad to believe he would one day grow up to change the world.

When Albert did start to talk, his parents
noticed he had a strange habit. He
would say things softly to himself,
sometimes over and over.

Then, finally,
he would say them out
loud to everyone else.

It seemed to Albert's parents that their son was
developing too slowly. They were so concerned that
they had a doctor examine him.

How would Albert learn anything in school if he had poor language skills?

One of his teachers would later make the comment that the boy probably wouldn't amount to much in life.

When Albert was about four or five, he got sick
and had to stay in bed for a few days.

Albert's father brought
him some toys to cheer
him up. One of them
was a compass.

A compass is a small
round device that looks
a little bit like a clock.
It has a needle attached to
the center that can spin around.

One end of the needle usually
has the head of an arrow,
and it always points in
a northern direction.

From the moment
Albert took the compass
in his small hands, he was
utterly fascinated by it.

Albert wondered why the compass needle always pointed north. He realized there were forces at work in the Universe that he could not see.

These forces affected the world and everything in it.

They were beyond the power of humans and had existed from the beginning of time.

So what was there in nature that made them behave the way they did?

Young Albert decided to devote all his time and energy to solving these mysteries.

Magical MAGNETISM

The force that makes a compass act the way it does is called magnetism.

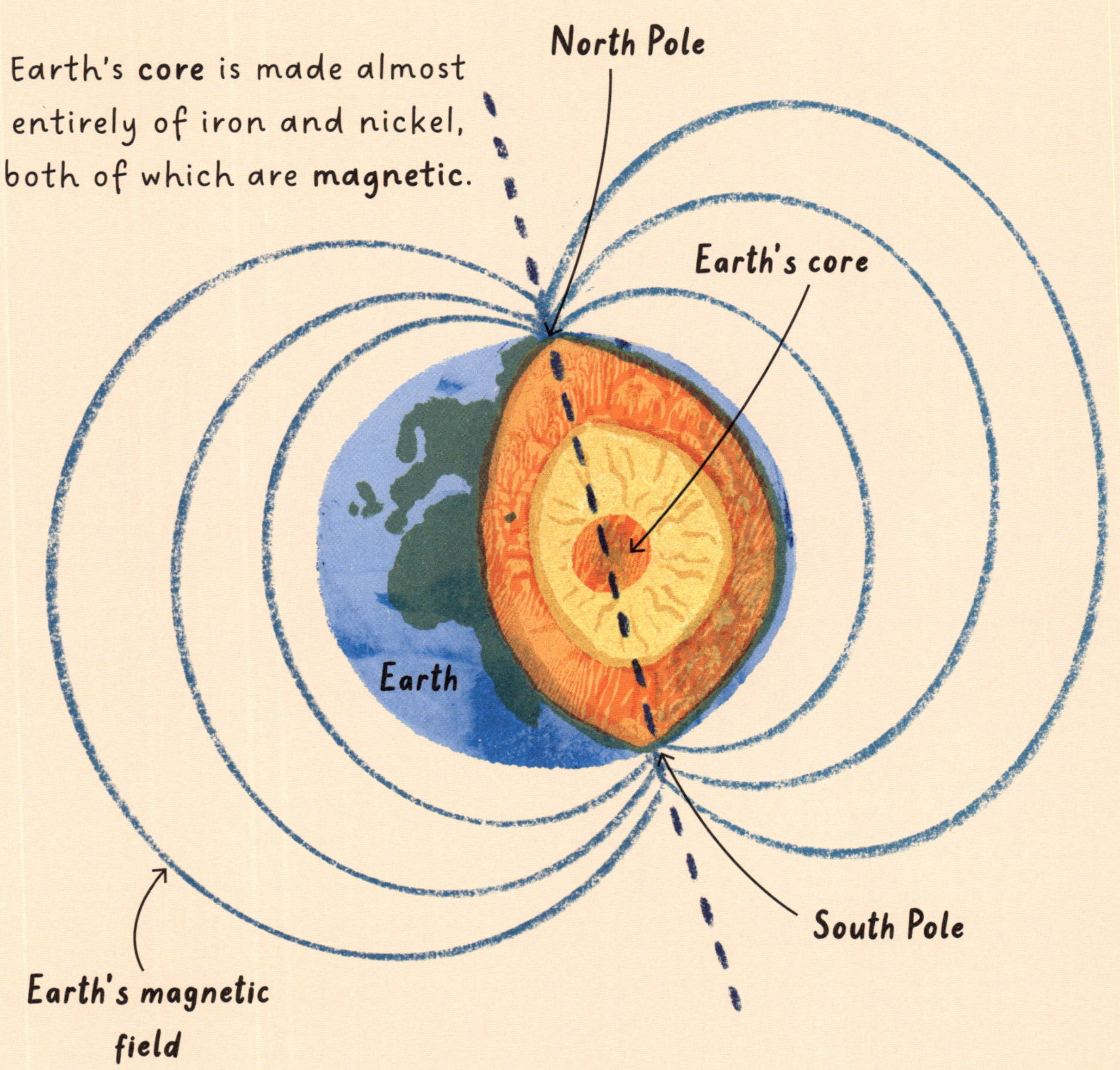

The metal in Earth's core creates a **magnetic field**—an invisible force—all around Earth. It is this force that acts upon the needle of a compass, which is a small **magnetized** piece of metal, making it point north.

Earth's **magnetic field** blocks some of the solar wind (harmful particles given out by the Sun), preventing it from reaching us.

The **needle** of a compass responds to the force of Earth's magnetic field by always pointing **North**.

People have used compasses to find their way around the world for centuries.

A FORCE TO BE RECKONED WITH

Physics Becomes Fascinating

Albert struggled with language difficulties during his first years at school, but that didn't mean he wasn't smart.

Though he wasn't very interested in grammar or other languages, he was intrigued by science and mathematics.

Back home, he would study alone for hours. This left little time for playing with friends, but Albert didn't mind.

His parents would buy books for him so he could stay well ahead of the other students. There were few things that got Albert more excited than new books.

Albert's parents were kind and generous people.
Sometimes they would help older students who didn't
have much money by inviting them over for dinner.

In 1889, when Albert was just ten years old,
Albert's parents invited a medical student
named Max Talmud to come to their home.
Max began having dinner with
the Einsteins on Thursdays.

It didn't take long for Max
to realize Albert was an
unusually smart little boy.

They would often talk about scientific subjects.
As a college student, Max's education was well
ahead of Albert's. And yet, Albert seemed to
have no trouble understanding Max.

Max began bringing books for Albert to read. They were not "kid's books" by any means. One was an illustrated science series with over twenty volumes.

The books that seemed to draw Albert's attention the most were about something called physics.

The physics books covered scientific principles about how objects and energy move. The reader was asked to picture experiments in order to understand complex ideas.

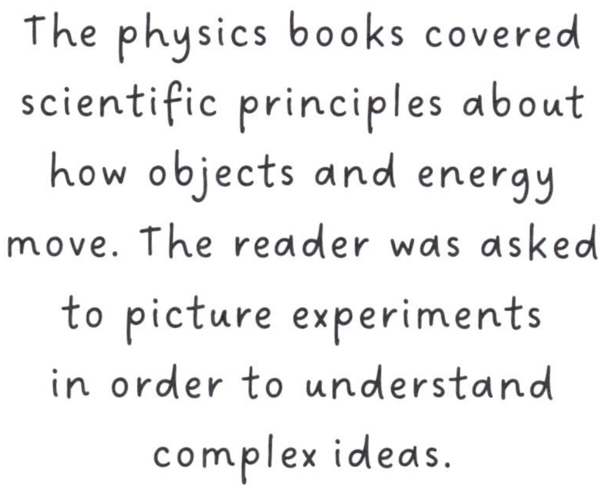

In one, you had to imagine yourself riding through outer space on an electrical signal!

When Max would show up for the next dinner, Albert would excitedly describe what he had learned the previous week.

Many of the books that Max brought had
exercises and experiments for the reader to do.

Albert would sit with a notebook and
work on these problems for hours.

He would lose track of time and
often forget to eat or bathe.

Then Albert would proudly show Max what he'd done the following Thursday.

One day, Max realized Albert's knowledge had passed even his own—in just a few months! He couldn't teach Albert any more science.

When it came to physics, Albert was already in a league of his own.

What is PHYSICS?

Physics is the study of how and why things move the way they do.

Physics involves all types of matter—which is anything that has weight and takes up space—and the study of forces that work upon matter.

Matter could be something very big, like an **airplane** . . .

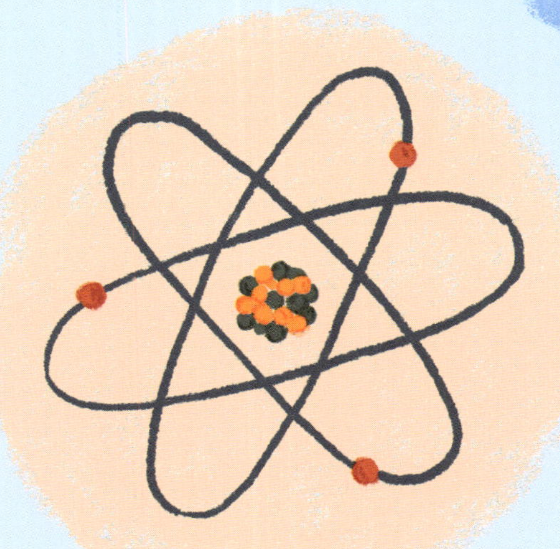

. . . or very small, like an **atom** (one of the tiny building blocks that make up everything around us).

Gravity, for example, is a powerful, invisible **force** that affects our lives every day.

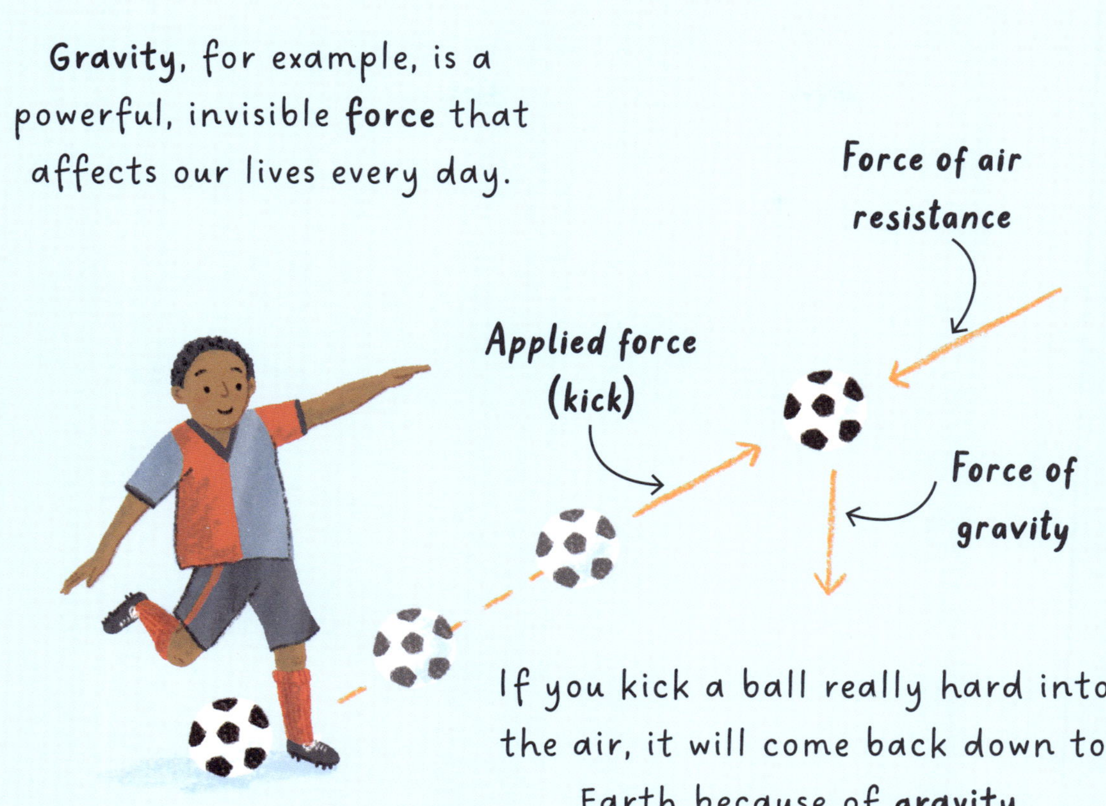

If you kick a ball really hard into the air, it will come back down to Earth because of **gravity**.

Physics also helps us understand the behavior of different types of **energy**, such as light and electricity.

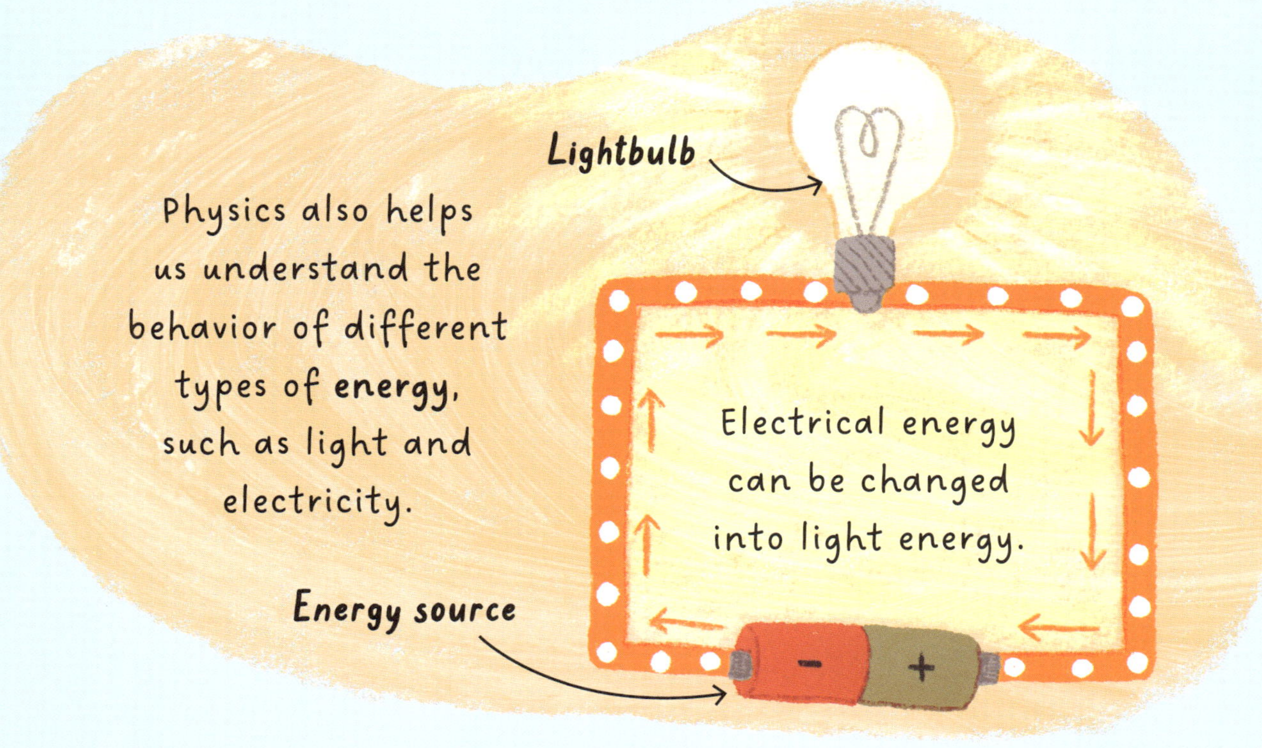

Albert would go on to make his greatest mark upon the world in the field of physics.

A REBEL IN THE MAKING
Science Needs Structure

As a teenager, Albert was well aware
of his brilliance. He was far ahead
of his classmates in science
and mathematics.

He became restless
and impatient, always
wanting to learn more
and take the next steps.

His parents understood all of his difficulties.
They wanted to make sure he got the kind of education
that would nurture his genius rather than hinder it.

Simply buying textbooks for him to read in
his spare time was no longer enough. Albert
was ready to move on to bigger things.

Albert's family was having serious money problems around this time. His father, Hermann, was forced to sell his electrical business in 1894 and move the family to Italy.

Albert, however, continued his schooling in Munich until the Christmas break. He then joined his family in Italy, but he didn't plan to stay for long.

He took the entrance exam to a school in
Switzerland called the Zurich Polytechnic.
This was a university, which meant he was
hoping to skip high school altogether!

Albert did very well on the
math and science parts of the
exam, but he failed in other
subjects, so he didn't get in.

Albert finished high school in 1896, at the age of 17. Then he was able to enroll in the Zurich Polytechnic.

It was famous for its courses in math, science, and engineering, so Albert was very excited.

But it wasn't long before he started causing some problems, too.

Albert had a habit of ignoring his professors. And sometimes when they would ask him to do something, he would do the opposite!

Albert began to think he was smarter than everyone else.

One of Albert's physics teachers, Jean Pernet, was in charge of a laboratory for running experiments.

Albert had a habit of showing up to Pernet's classes only when he felt like it.

There are specific ways that experiments are supposed to be done. But Albert did them only in the way he thought best.

Then Albert's luck ran out. In the summer of 1899, he was doing an experiment in Pernet's lab and caused an explosion.

He badly injured his right hand and was unable to use it for weeks.

Albert learned that his brilliance didn't mean he could ignore the rules—success in science required discipline, too.

The Scientific METHOD

There is a right way to conduct scientific experiments. It is known as the "scientific method."

You start with a clear understanding of the **problem** you're trying to solve.

1. Your first step is to make **observations** and collect whatever data you can.

2. From there, you make a reasonable guess about the answers you're seeking. This is called the "**hypothesis**."

3. Next, you run a series of **experiments**—tests that will help support your guess or make you want to rethink it.

4. Once the experiments are completed, you have to review the **results** (the information you've learned).

5. Finally, you draw your **conclusions**.

This highly effective step-by-step approach has served scientists very well for centuries.

LIGHT FANTASTIC
Waves and Particles

In 1900, Albert graduated from the Zurich Polytechnic— but only just. Despite his cleverness, he was ranked fourth in a class of five.

This meant he had some of the lowest grades among his fellow students.

Part of the reason for this probably had to do with his habit of making his professors angry.

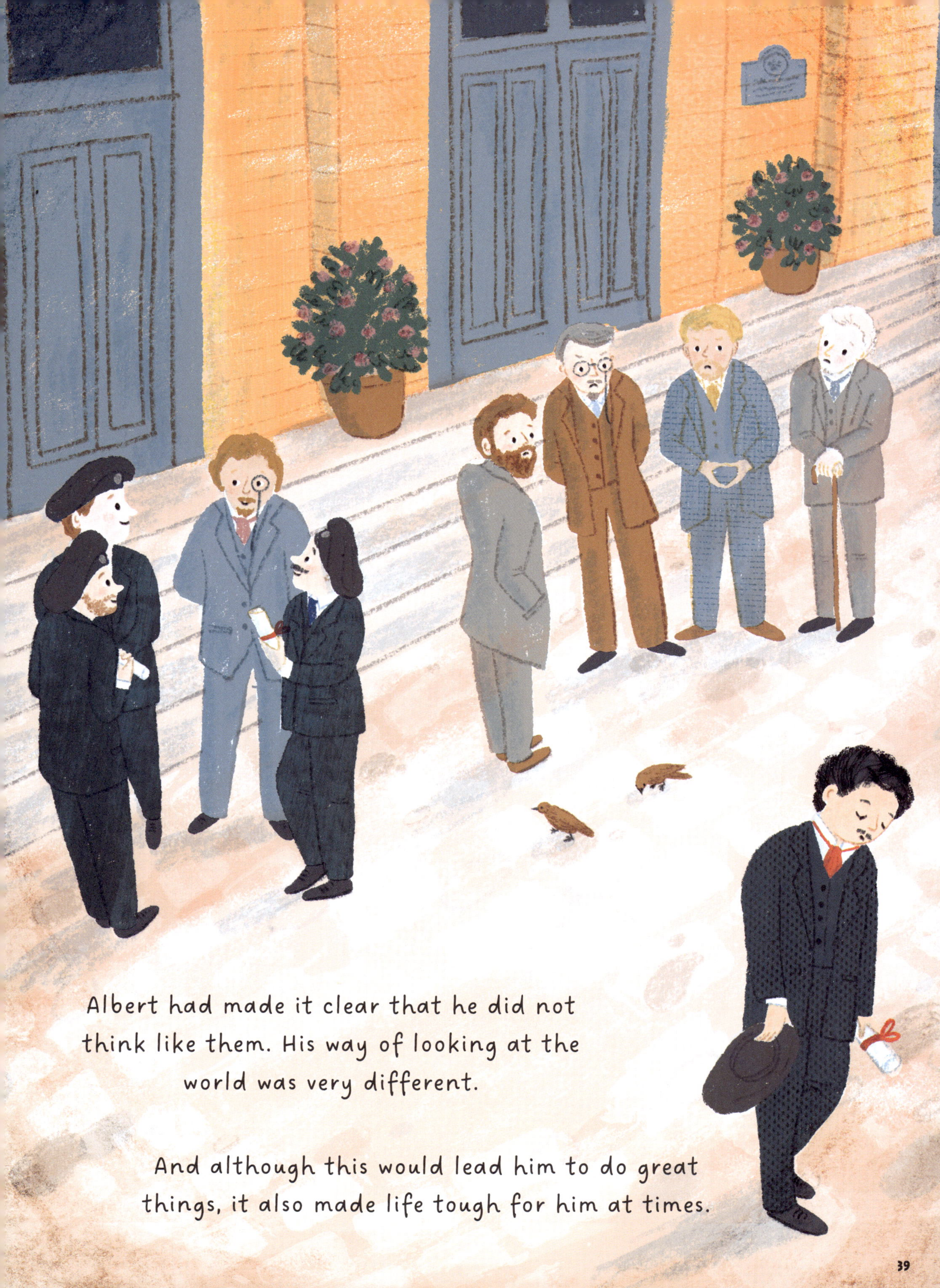

Albert had made it clear that he did not think like them. His way of looking at the world was very different.

And although this would lead him to do great things, it also made life tough for him at times.

Albert had met someone—a fellow student—at the Zurich Polytechnic. She was named Mileva Marić, and she and Albert fell in love.

After he graduated, Albert traveled to join his family, who were on vacation in the Swiss mountains. They were happy to be reunited.

Albert knew he would
have to get a job if
he wanted to marry
Mileva and support
a household.

But he also had a
few ideas about physics
that he wanted to
share with the world.

The following year, 1901, Albert became a Swiss citizen. He had been born a German citizen, but he had given up his German citizenship back in 1896.

He didn't like the way Germany forced its male citizens to join the military. He felt Switzerland was kinder and gentler toward its people.

Then Albert tried to find a teaching job. He did become a tutor for awhile. A tutor is a private teacher who works with pupils individually.

And in his spare time, one of the ideas Albert was developing was called the photoelectric effect. This attempts to explain what happens when light shines on certain things, specifically metal.

The photoelectric effect had first been noted in 1887 by a German physicist named Heinrich Hertz.

Hertz and other physicists came to believe that light traveled through the air as a wave.

But Albert wasn't so sure about this, so he started to study the concept in greater detail.

Albert was on the right track! He was the first person to understand that light is not just a wave, but both a wave and a group of particles.

His paper was published on June 9, 1905. Later experiments confirmed Albert's calculations were correct.

Albert's idea changed our understanding of light and energy.

The Photoelectric EFFECT

The photoelectric effect proves that light is made up of individual particles.

It happens when light **particles** strike a piece of metal.

These light particles are called **photons**.

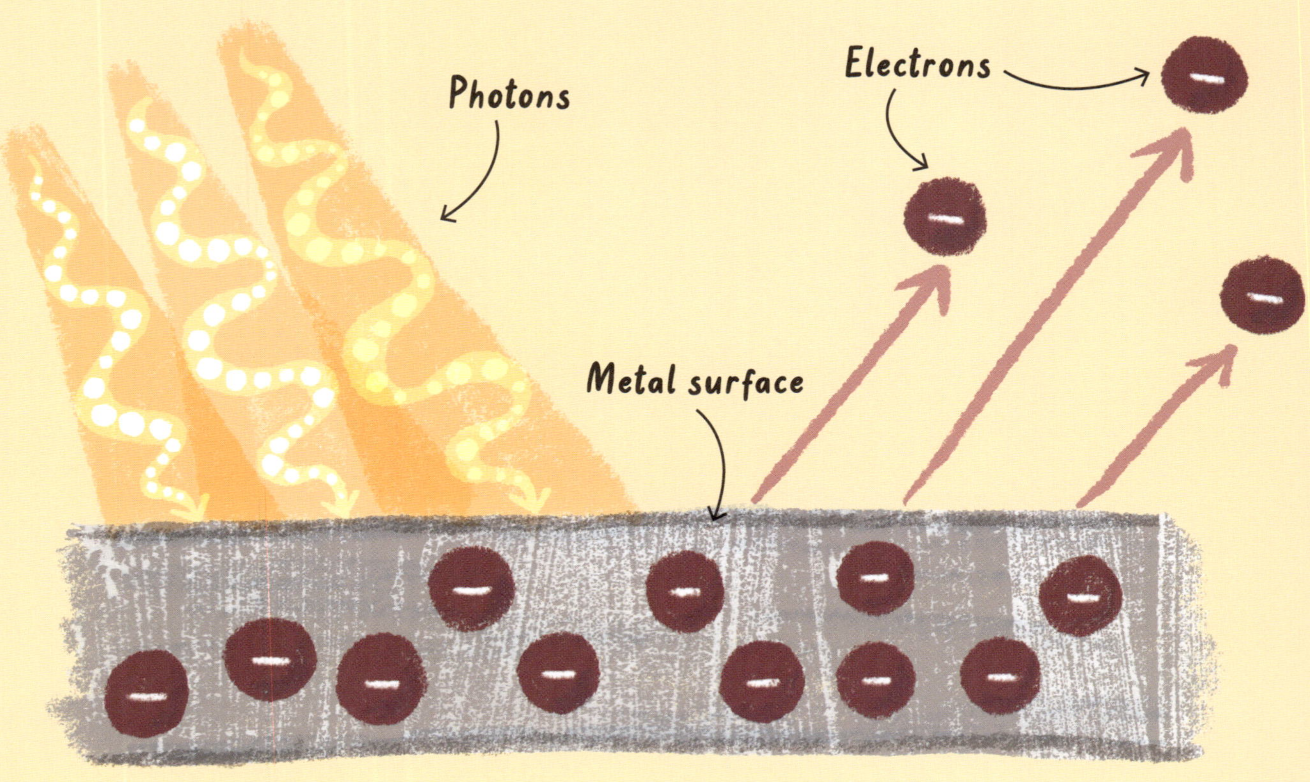

When photons strike a metal surface, particles in the metal are jarred loose and thrown off. Those particles are called **electrons**.

Today, there are many uses for the photoelectric effect.

A **solar panel**, for example, uses the light from the Sun to create electricity.

Digital cameras also use the photoelectric effect to convert light into electrical signals.

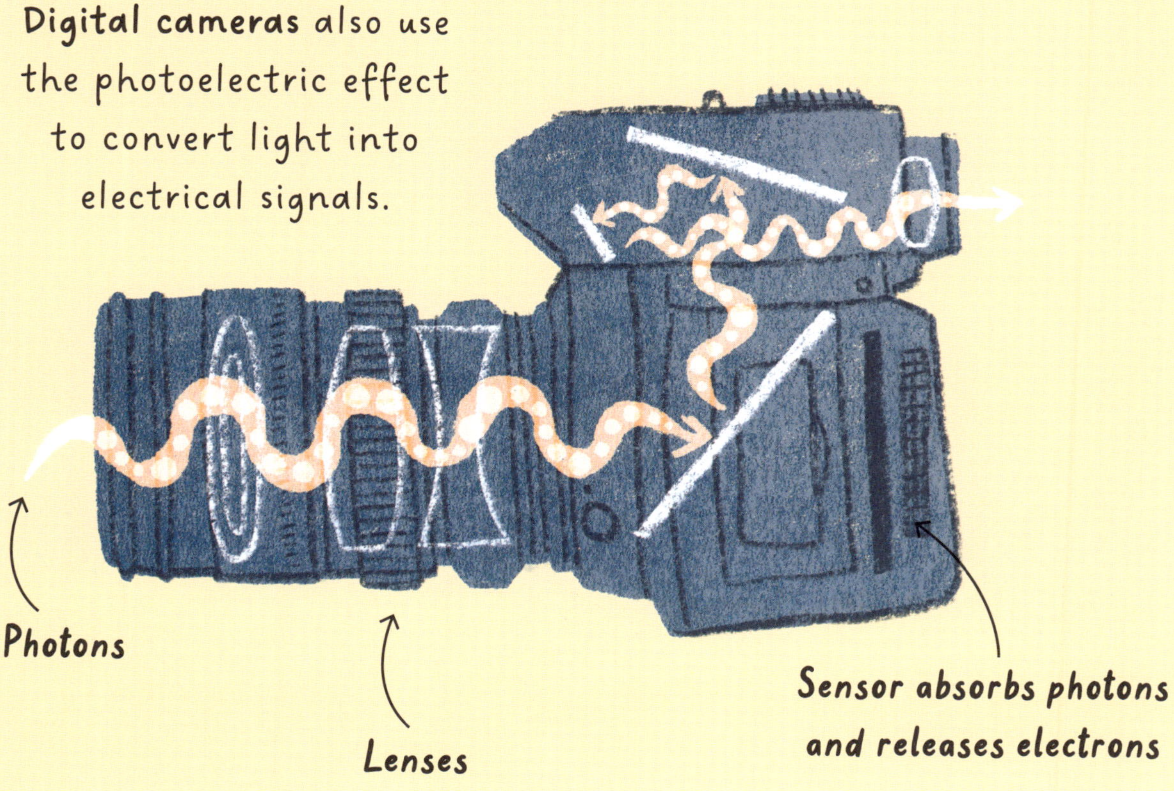

MOTION COMMOTION
Friendships Forge a Theory of Forces

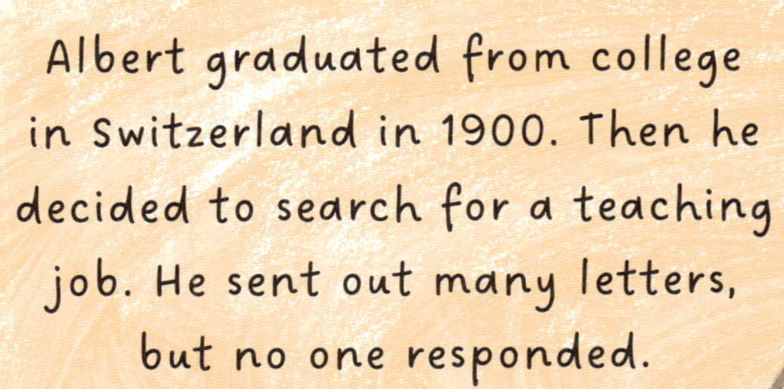

Albert graduated from college in Switzerland in 1900. Then he decided to search for a teaching job. He sent out many letters, but no one responded.

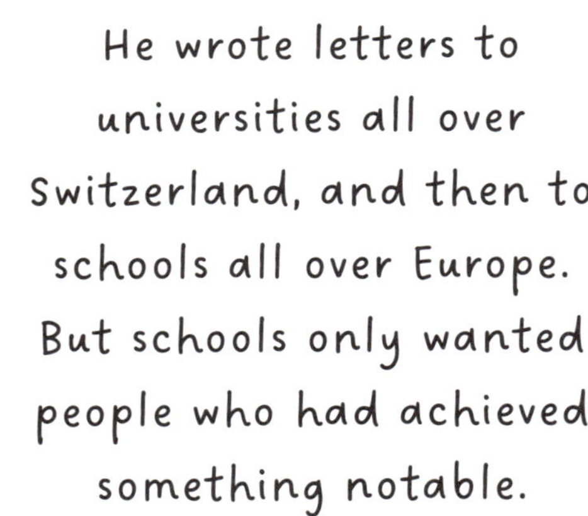

He wrote letters to universities all over Switzerland, and then to schools all over Europe. But schools only wanted people who had achieved something notable.

Albert got so desperate that he even went back to the university he had attended, the Zurich Polytechnic, to see if someone would hire him.

But there was no chance of that happening. He had been a very difficult student for many of his old professors, so they turned him away.

Albert thought it would help if he wrote an article about physics and had it published.

There were some very good scientific journals at the time. They featured articles on advanced scientific subjects.

One was called the Annals of Physics, which Albert admired very much.

Albert wrote an article about the capillary effect. This describes the way liquid moves against gravity through a tight space. A good example of this is water moving up through a straw.

Albert didn't think his article was very good, but the *Annals of Physics* published it in 1901.

Albert's relationship with his girlfriend, Mileva, was becoming serious. He knew he wanted to build a life with her.

But his "day job" as a tutor was not making enough money.

So he placed an ad in the local newspaper to offer his tutoring services to other people.

One of the first people to answer the ad was
a student named Maurice Solovine.

Then another young man, a mathematician
named Conrad Habicht, joined them.

Albert never actually tutored
them, so he didn't make any
extra money! He did, however,
become very good friends with
both Maurice and Conrad.

Albert, Maurice, and Conrad would spend long hours talking. They usually met in Albert's apartment, and even gave themselves a name—the Olympia Academy.

Maurice and Conrad likely helped Albert with some of his new ideas in physics. He had one around this time about something called Brownian motion.

This was a way of understanding how particles moved within either water or gas.

It was first described by a botanist named Robert Brown in 1827. Brown saw grains of pollen through a microscope, moving in water.

Brown thought maybe the movements were entirely random, but Albert had another theory.

Albert decided that the force behind Brownian motion was the movements of tiny particles such as atoms.

Brownian MOTION

Einstein realized that the atoms and molecules causing Brownian motion had an energy all of their own, so they were always moving.

When other matter came in contact with the **moving** atoms and molecules, those things also moved.

Some scientists had suggested the presence of atoms and molecules in the past. Albert, however, was the first to **prove** they were really there.

Today, Albert's concepts for Brownian motion are used to try and predict other seemingly **random** events, such as the future of the stock market.

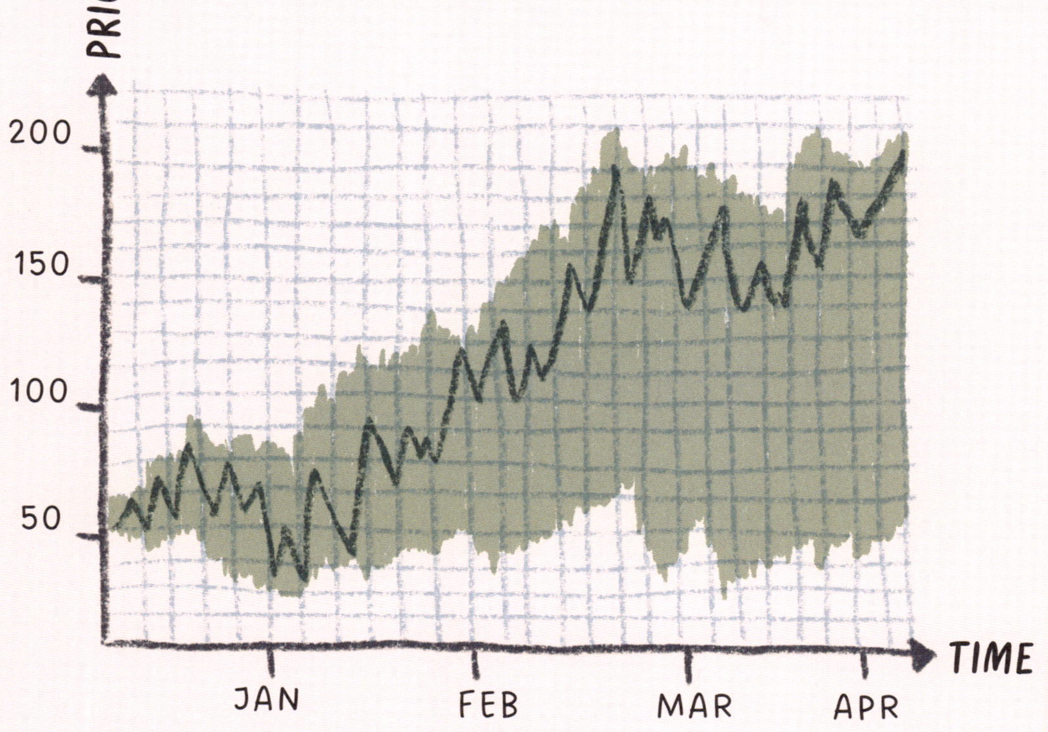

They are also helpful in the growing field of **nanotechnology**.

For example, tiny **nanoparticles** are now commonly used to create water-repelling fabrics.

ANNALS OF PHYSICS
A Very Special Theory Takes Shape

Albert graduated from college in the spring of 1900—and now he needed a job.

He had many amazing ideas about physics that he loved to discuss with his friends. But that wasn't earning him any money.

Then Albert's friend Marcel Grossman told him there was
a position available in a Swiss patent office.

A patent is issued
by a government
to recognize an
invention that
someone created.

Albert decided to
apply. He and his
fiancée, Mileva, could
not have a secure
future together until
he could support
them both.

Albert started working at the patent office in June of 1902. The office building was located in the Swiss town of Bern. He started as a third-level patent clerk—the lowest rank.

Albert's job was to review patent applications. Part of this meant figuring out if the inventions worked correctly. Lots of people wanted to get a patent for their inventions, but not all of them worked!

He also had to determine if an application was too similar to an existing one. You could not get a patent for something that had already been invented.

This meant checking new applications against older designs that had already been registered.

Albert told friends that the job in the patent office was boring at times. But he enjoyed the quiet, and he didn't mind spending long hours by himself.

It gave him a chance to further develop his physics ideas.

Albert was captivated by the relationship between space, time, and energy, and he was beginning to draw some interesting conclusions.

He was also able to support Mileva, and they got married in January of 1903.

Away from the office, Albert would meet friends to discuss his ideas. He loved music, and playing his violin helped him think through tricky problems.

Crucially, Albert had time to further develop a key principle of physics—one that had first been described by the famous scientist Galileo Galilei back in 1632.

In the past, scientists believed that space and time were separate and unrelated. In other words, general principles that applied to each did not apply to both because there was no connection between them.

Albert's theory of "special relativity" showed the world that space and time are, in fact, very much related to one another.

He also realized that the speed of light is a true constant in the Universe. This means it moves at the same speed no matter how fast anything else is moving.

Special relativity revolutionized our understanding of space and time.

Special RELATIVITY

Einstein's theory of special relativity explained that observed motion varies, depending on viewpoint.

Imagine you are on a train traveling at 50 miles per hour and throw a ball at a speed of 5 miles per hour. **From your viewpoint** the ball is moving independently, at 5 miles per hour.

Now imagine someone outside the train. **From their viewpoint**, you and the train are both moving at 50 miles per hour, and the ball is moving at 55 miles per hour.

The ball's speed is relative—it depends on the viewpoint of the observer.

Albert's theory also revealed that **light speed** is the top speed achievable in the Universe—nothing can travel faster.

The speed of light is **186,282 miles per second,** or 299,792,458 meters per second.

It takes over **8 minutes** for light to travel from the Sun to Earth, depending on the time of year.

Albert stated his theory mathematically in an **equation:**

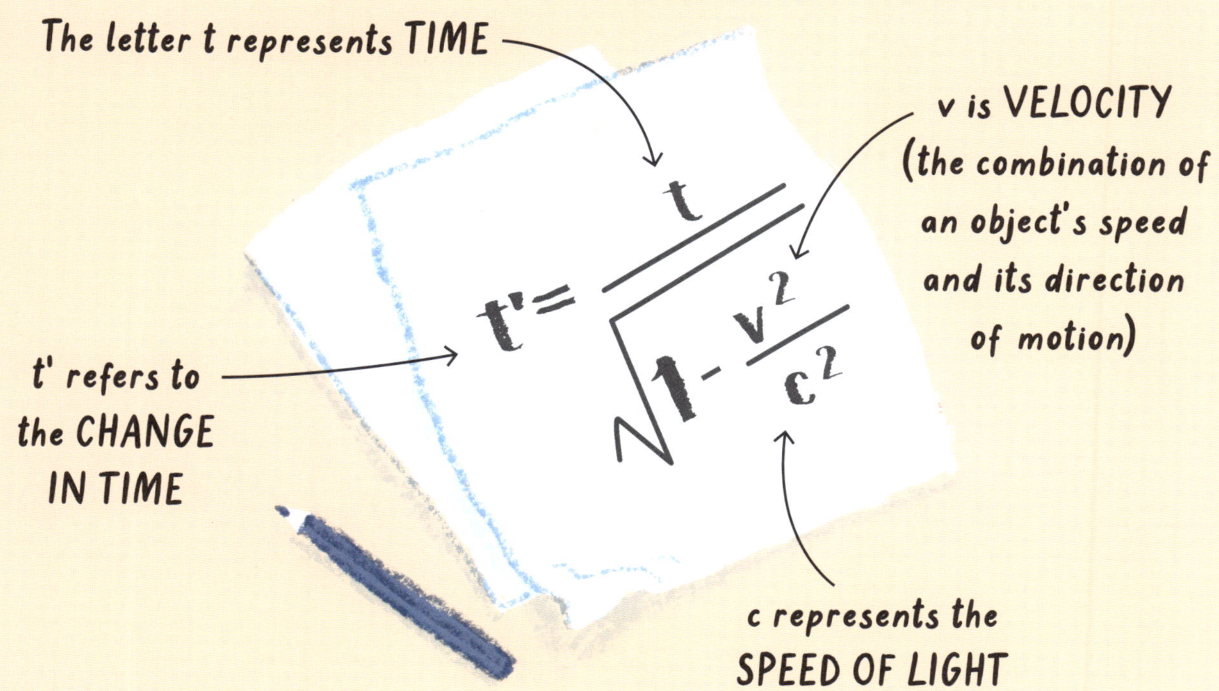

$$t' = \frac{t}{\sqrt{1 - \frac{v^2}{c^2}}}$$

THE MOST FAMOUS EQUATION
A Scientific Principle is Overturned

Many exciting things were happening to Albert in the early 1900s. His first son, Hans, was born on May 14, 1904.

Albert's wife, Mileva, was worried that Albert might not be too happy about being a father, as he was very busy.

But Albert seemed thrilled by Hans. He would build toys for him. Many years later, Hans remembered a cable car that Albert made from a matchbox and some string.

Although Albert set aside time to be playful with Hans, he was also spending many hours developing new ideas in physics.

In his job at the patent office, Albert would often get a day's work done in just a few hours. Then he'd spend the rest of the time working on his physics ideas.

There were piles of papers all over his desk. Some had to do with the patents. But many more were his physics notes. He would hide them whenever anyone came by!

His boss was a man named Friedrich Haller. Friedrich knew what Albert was doing, but he never said anything.

Friedrich realized that Albert was on to something and needed time to work.

The ideas behind basic physics had been around for centuries. Many of them came from an English physicist and mathematician named Isaac Newton back in the late 1600s.

Very few ideas concerning physics had changed in the centuries after that.

As the 18th century drew to a close, a famous British scientist named Lord Kelvin even said there was nothing new left to discover in physics.

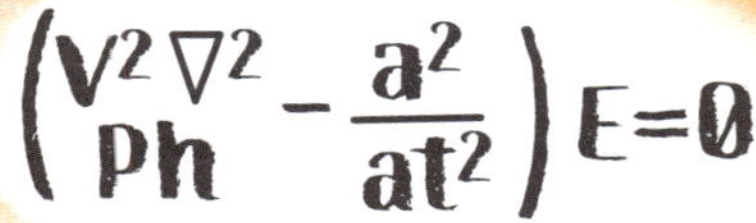

This was exactly the kind
of closed-minded thinking
Albert did not like.

He had been working on some new and
exciting ideas for a long time. Now he felt
most of those ideas were ready to share.
The days of keeping them to himself were over.

When Albert was ready to present his new physics ideas to the world, he started writing articles. His plan was to publish each one in the scientific journal *Annals of Physics*.

He had no idea he was about to forever change the way humans looked at just about everything.

Albert was willing to question anything that didn't
make sense, even established scientific theories.

One of Albert's ideas had
to do with something
called mass-energy
equivalence.

He believed that
everything that had
mass (physical form) also
contained some energy.
And that each could be
changed into the other.

To explain this, Albert came up with
what is probably the most famous
equation in history: $E=mc^2$.

Albert's mass-energy equivalence concept suggested that all things with mass contained some energy.

Matter refers to anything that has physical qualities. A chair, a table, a person . . . they are all matter. **Mass** is a measure of the amount of matter in an object.

Because the speed of light is such a big number— about 186,282 miles per second—Albert's equation supported his belief that even the **smallest amount** of mass could contain a **huge amount** of energy.

Albert's theory said that mass and energy were basically the **same thing**, but in different forms . . .

The concept is used today in a reaction called **nuclear fission**, which we use to create nuclear power. In nuclear fission, an atom's **nucleus** (tiny central part) is split into smaller "nuclei," releasing energy.

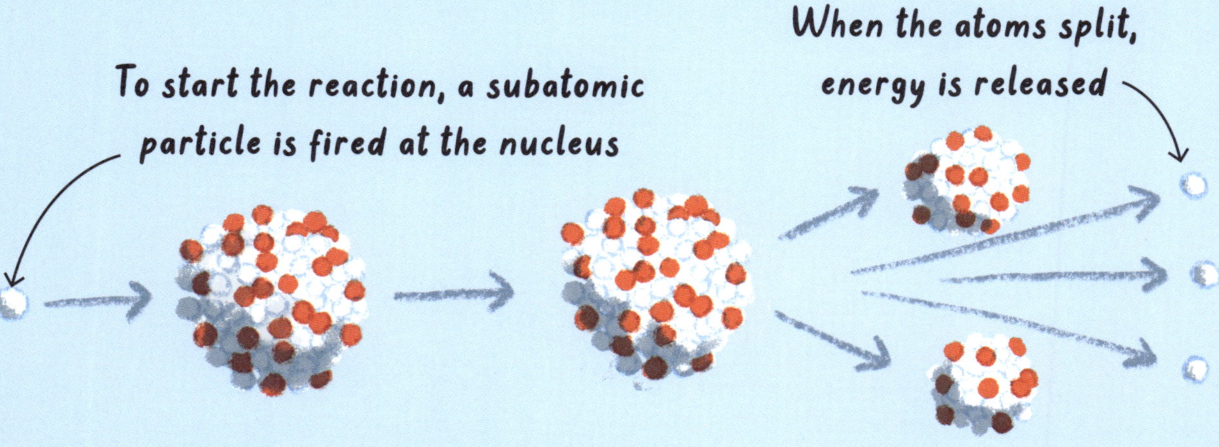

Annals of Physics received Albert's paper on September 27, 1905, and published it two months later, on 21 November, 1905.

THE MIRACLE YEAR
Turning Gravity Upside Down

The year 1905 would become known as Albert's "miracle year." He finished working on some of his most brilliant new ideas in physics, and wrote articles about each of them.

There were four articles in total. Each solved a problem that had been puzzling physicists for centuries.

He published the four articles in the scientific journal Annals of Physics. Other physicists around the world read these articles and were very impressed.

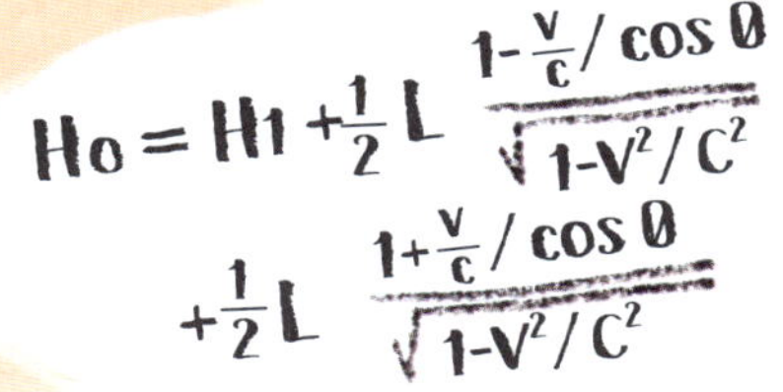

$= H_1 + \dfrac{L}{\sqrt{1-V^2/C^2}}$

To many academics, Albert was a complete unknown. They were very interested in this man who was shedding new light on long-established theories.

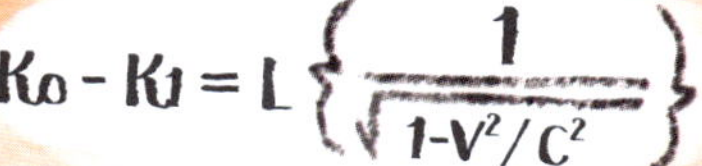

Albert's plan was to have a job in a university that had a physics department. He felt he could get bright young students excited about physics and his new ideas.

He'd had a job reviewing applications for new inventions since 1902, and it had earned him a decent living. But after his four articles were published in *Annals of Physics*, he was ready to move on.

In 1908, Albert was offered his
first teaching position, at the University
of Bern. A year later, he began working
at the University of Zurich.

But then, in 1913, Albert had the chance to set
aside teaching and work only on his physics ideas.

Albert's marriage had been unhappy for a while.
So instead of going with Albert to Germany, his
wife Mileva and their two sons moved to Zurich.

Albert found himself an apartment in the city of Berlin
in April of 1914. He and Mileva divorced a few years later.

Just a few months after Albert arrived in Berlin,
World War I broke out, when Germany declared
war on Russia and France.

Albert did not like
Germany's involvement
in the war. But he could
not be distracted
from his theories.

He had another great one to share with
the world—and it was all about gravity.

For centuries, scientists believed that gravity was nothing more than a force of attraction between two objects. This came from English scientist Isaac Newton, who lived from 1643–1727.

Isaac was a hero of Albert's—later in his life he called him a "shining spirit." But he thought that Isaac's ideas about gravity weren't quite correct.

Albert had been running experiments to
support his own ideas since the mid-1900s.
For example, he wondered how the force
of gravity worked while he stood in
a moving elevator.

By 1915, Albert
believed he had
proof that his new
ideas were right. He
wrote another article
for the Annals of
Physics in 1916—and
once again changed
the world.

Albert's new approach to gravity
became known as the general
theory of relativity.

Albert's theory proved that gravity isn't simply a force that pulls things down.

Instead, massive objects **warp** or curve both the space and time that surround them.

Another key part of Albert's theory is that space and time are not separate. They are woven together into something called "**spacetime.**"

If a large rock is placed into the center of a trampoline, the trampoline dips down at that point. That dip is a bit like how the force of a massive object causes a **warp in spacetime**.

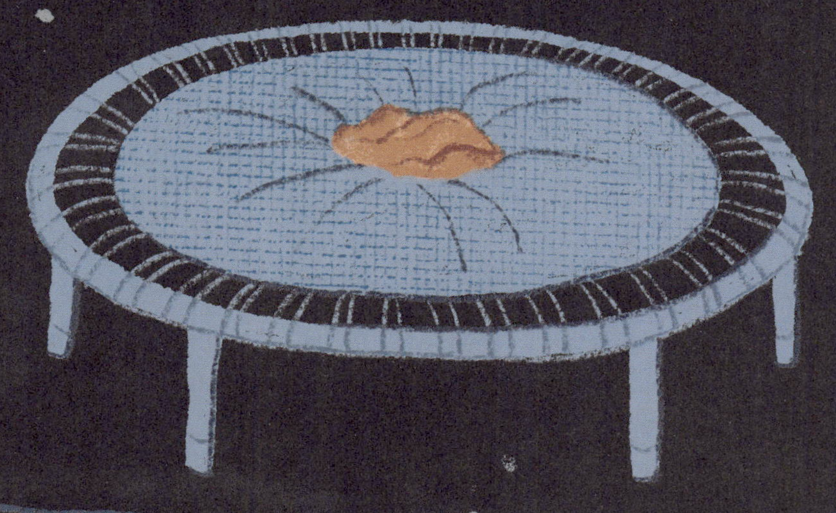

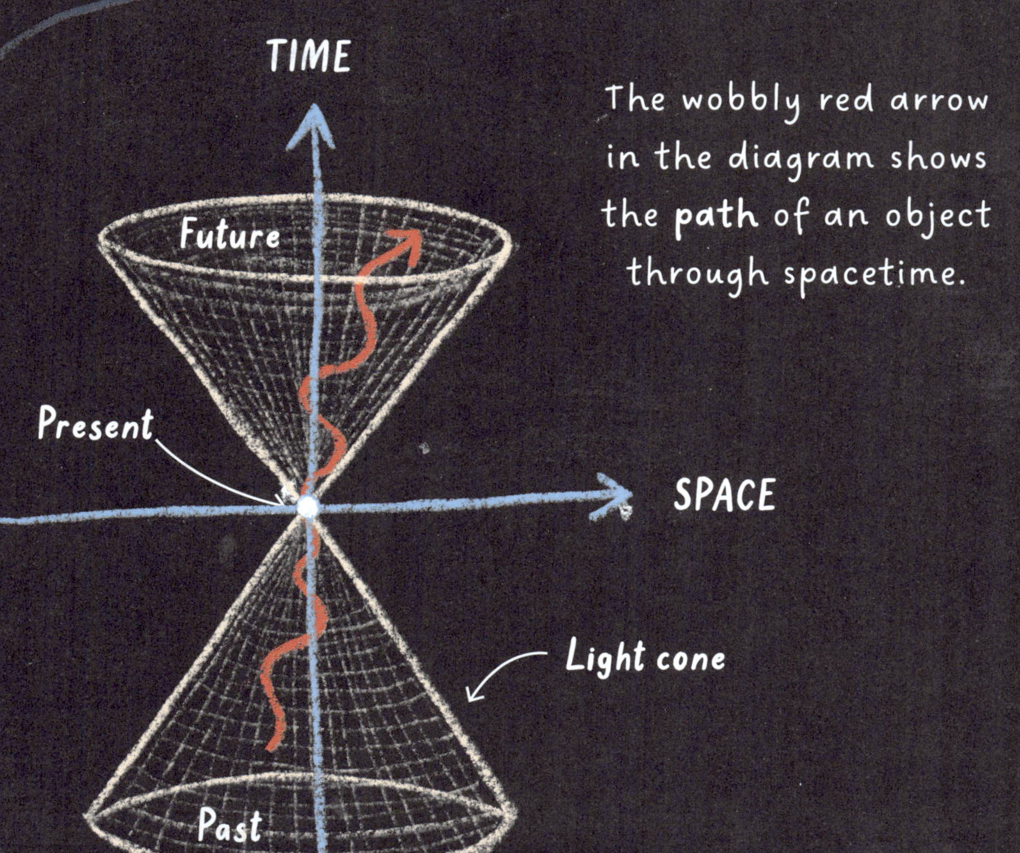

The wobbly red arrow in the diagram shows the **path** of an object through spacetime.

THE BIGGEST PRIZE OF ALL
Proof at Last for a Famous Theory

Albert's 1916 article concerning his general theory of relativity changed the way people thought about gravity. He said later in his life that he considered it his greatest accomplishment.

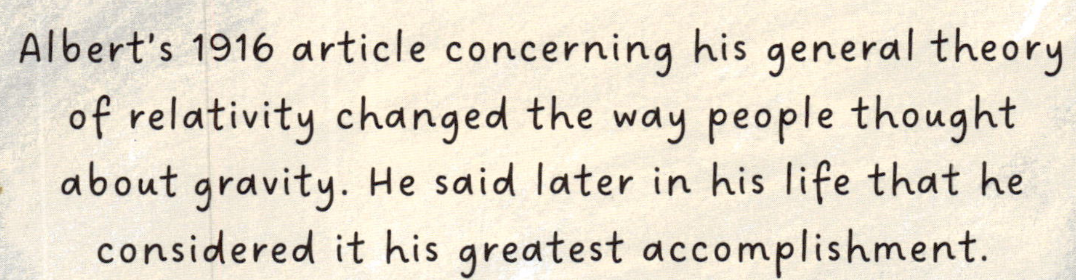

Albert wrote in the article that gravity was not merely some force between two objects. Instead massive objects affected both the space and time that surrounds them.

Scientists everywhere began talking about this amazing idea.

It was the first new theory on gravity in centuries, and it explained several phenomena that had been puzzling scientists for years.

The theory accounted for the speed of the planet Mercury's precession (the slow wobble it does as it spins)—a mystery that astronomers had long been trying to solve.

But other physicists around the world weren't ready to accept this new idea just yet. Explanations were not the same as *proof*.

In 1919, scientists were given a fantastic opportunity to test Albert's theory.

There was a solar eclipse on May 29th. A solar eclipse happens when the Moon moves in front of the Sun, blocking its light from reaching Earth.

This gave scientists a chance to see stars that were normally hidden by the Sun's bright glare.

The scientists took pictures of those stars, then more pictures after the Sun had moved away. Then they compared the pictures and made an amazing discovery.

Those stars near the Sun did appear to have moved a little bit. This meant the light that came from those stars had been bent by the Sun's gravity. Albert's theory was correct!

The news that Albert's new theory on gravity was correct became a big story all over the world—and not just within the scientific community. It was being reported on the radio and printed in thousands of newspapers.

All of a sudden, Albert was famous. In fact, he was considered the very first celebrity scientist.

He came to America in April of 1921 to do a tour. There were huge parties thrown in his honor.

He also visited some of
the best colleges in the
country to give talks
about his discoveries.

Accompanied by
his second wife, Elsa,
he even got to visit
the White House!

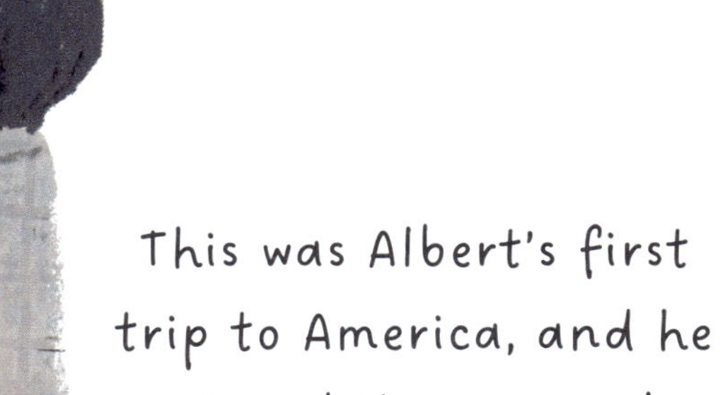

This was Albert's first
trip to America, and he
enjoyed it very much.

In September of 1922, Albert received some amazing
news. He was being awarded a Nobel Prize.

The Nobel is given each year to those who
have made outstanding achievements in
one of a few different categories.

These categories are physics, chemistry,
literature, peace, and physiology
or medicine.

The group of scientists who gave him the award hoped Albert
would attend the awards ceremony in Sweden in December—
but the timing interfered with other travel plans.

Albert had also been turned down for the Nobel Prize in past years. He was pretty annoyed about this, so he decided not to show up to receive it.

Besides, his genius was now being celebrated on a much wider stage than just within the world of science.

Albert's hard work finally paid off—his brilliance was being recognized.

A Legacy of IDEAS

Albert's world-renowned theories inspired a new generation of physicists, many of whom would build on Albert's ideas in the years to come.

Kip Thorne was awarded the Nobel Prize for Physics in 2017 for his work observing gravitational waves. These waves are ripples in **spacetime**.

Andrea Ghez has spent decades studying black holes as they relate to Albert's Theory of General Relativity.

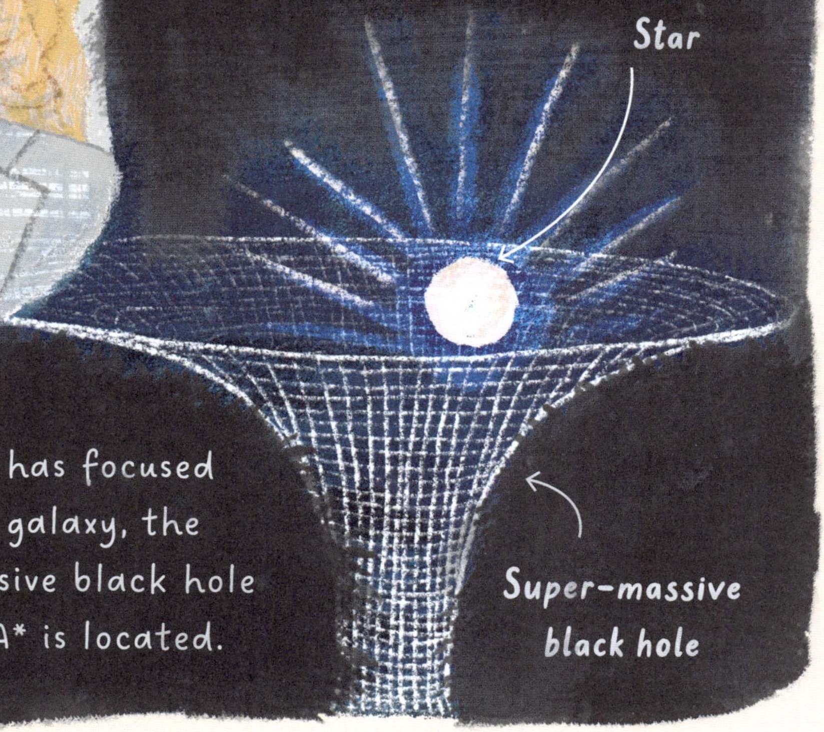

Much of her research has focused on the center of our galaxy, the Milky Way, where a massive black hole known as Sagittarius A* is located.

Stephen Hawking used some of Albert's ideas to improve our understanding of black holes and the **origins of the Universe.**

The afterglow of the birth of the Universe can still be detected as radiation.

ONE MORE BIG IDEA
Can Everything Be Explained?

Albert kept very busy throughout the later years of his life.
He had been born in Germany and lived there for a long time.
But then Adolf Hitler's Nazi Party rose to power in the 1930s.

The Nazis believed Jewish
people were inferior. As a famous
scientist who was also Jewish, Albert
was becoming unsafe in Germany.

So he moved to America and began teaching
at Princeton University in New Jersey.

He was given offers to teach in other countries as well.
But he decided to stay in America for good in 1935.

He liked that a person could speak
their mind in America without the risk
of getting punished by the government.

Once Albert became settled in Princeton, he got
busy working on new ideas in physics.

He had already come up with some of the most
important physics theories in history.
But he always had more.

Albert worked with physicist Nathan Rosen on the
Einstein-Rosen Bridge Theory. It suggested that there
were "wormholes" in space—tunnels that allowed for
travel over great distances in a relatively short time.

Albert also helped develop
a field of study known as
quantum mechanics, which
tries to better understand
how and why particles such as
atoms act the way they do.

He described the strange connections
between some atoms, and suggested that
there was much more to be discovered.

While he was in America, Albert continued working on
an idea that he had started developing in the 1920s.

Albert's goal was to create a
single set of rules that would
explain all the forces that
controlled the Universe.

This would include such things as
gravity, electromagnetism, binding,
light phenomena, and more.

Albert had long believed that there was a connection
between all things in the known world.

Now he wanted to
join them together in
one elegant and simple
explanation for everything.

This ambitious concept came to be
known as the unified field theory.

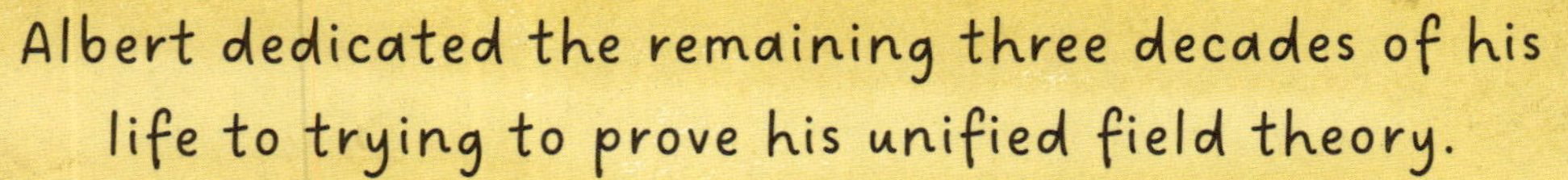

Albert dedicated the remaining three decades of his
life to trying to prove his unified field theory.

While he approached the problem many times
from different angles, he never perfected it.

When he died on April 18, 1955,
he still had thousands of pages
of notes, but no solution.

He was one of the greatest geniuses in human history.
And yet there were some things even he couldn't accomplish.

Other physicists have continued
working on Albert's idea. It has greatly
expanded the field of theoretical physics,
where ideas are explored in detail.

Unified field theory may never be proven,
but continued work on it has contributed to
other scientific breakthroughs.

What is COSMOLOGY?

The study of the Universe, including how it works, what it contains, and its long history, is called cosmology.

Albert's theories help explain the **movements** of massive objects, such as the paths of planets orbiting around stars.

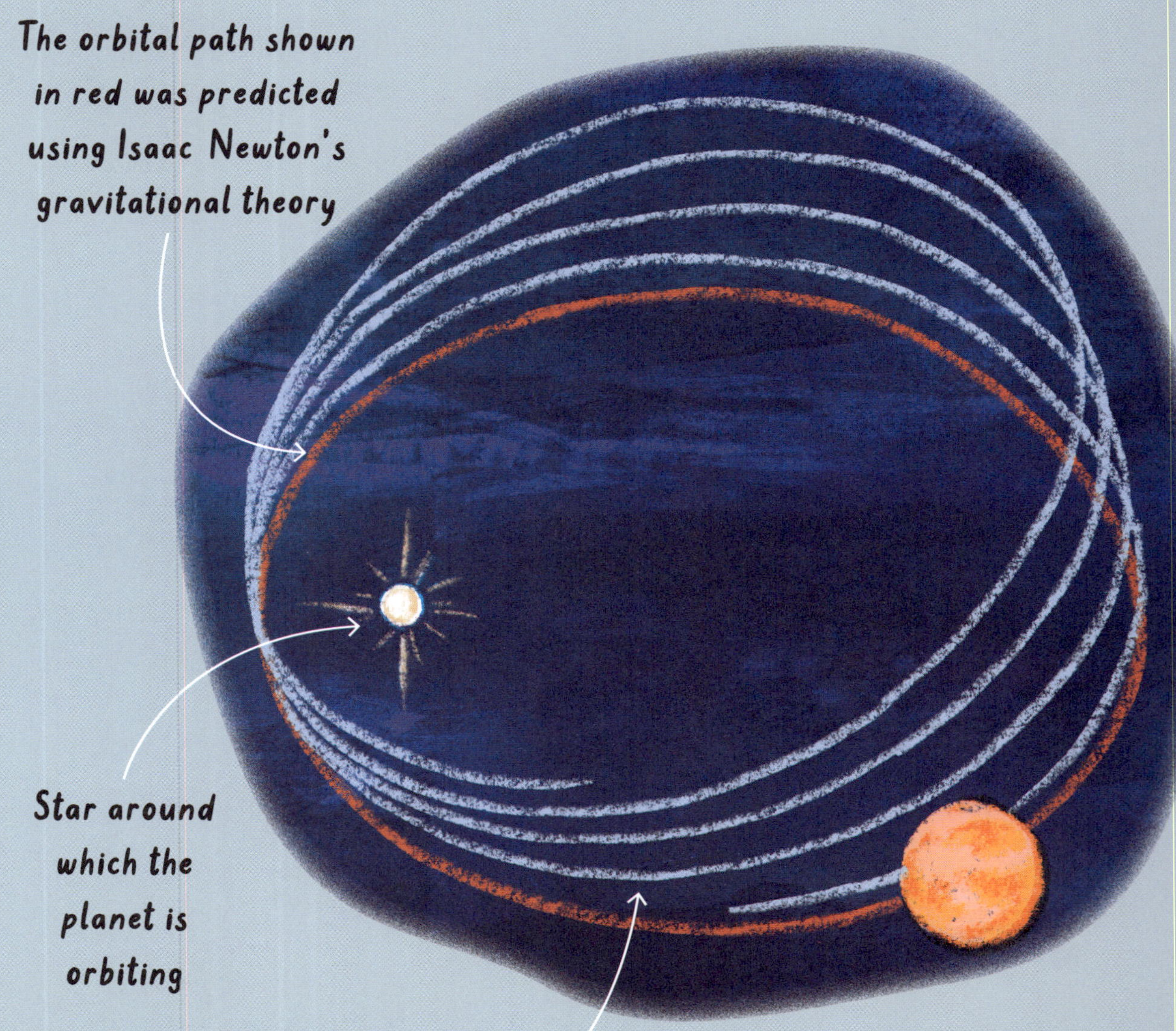

Albert's ideas have also contributed to theories about the **origins** of the Universe, and led us to the idea that the Universe is always growing.

1. Birth of the Universe
2. First particles
3. First light
4. First stars
5. First galaxies
6. Today

ALBERT EINSTEIN'S JOURNEY TO BECOMING A GROUNDBREAKING PHYSICIST

1879

March 14

Albert Einstein is **born** in Ulm, Germany.

1889

Albert meets Max Talmud, a friend of his parents, who gives Albert a set of **science books**.

1895

Albert takes the entrance exam for the **Zurich Polytechnic**. He excels at the math and physics parts, but fails the rest.

1905

November

The last of Albert's *Annus mirabilis* ("Miracle Year") papers is published. It is about **Mass-Energy Equivalence**.

1905

September

The third paper, containing the theory of **Special Relativity,** is published.

1911

Albert becomes a full **professor** of physics.

1916

Albert publishes a paper covering another groundbreaking theory, that of **General Relativity**.

1955

April 18

Albert dies at the age of 76.

1935

Albert publishes a paper with Nathan Rosen about connections through spacetime that will become known as "**wormholes**."

1896

Albert enrolls at the **Federal Polytechnic School** in Zurich.

1901

Albert becomes a **Swiss** citizen.

1903

Albert marries Mileva Maric. They met as fellow students at university.

1905
July

Albert's second paper of the year, on **Brownian Motion**, is published.

1905
June

The first of four of Albert's scientific papers that **forever alter the field of physics** is published. It is on the **photoelectric effect.**

1919

Albert and Mileva divorce, and Albert marries **Elsa Löwenthal**.

1921

Albert tours the United States and becomes the world's first **celebrity scientist**.

1933

Albert moves to the United States and settles in **Princeton**, New Jersey.

1922

Albert is awarded the **Nobel Prize** for Physics.